HOLZ books | 2018

# CHEMISTRY

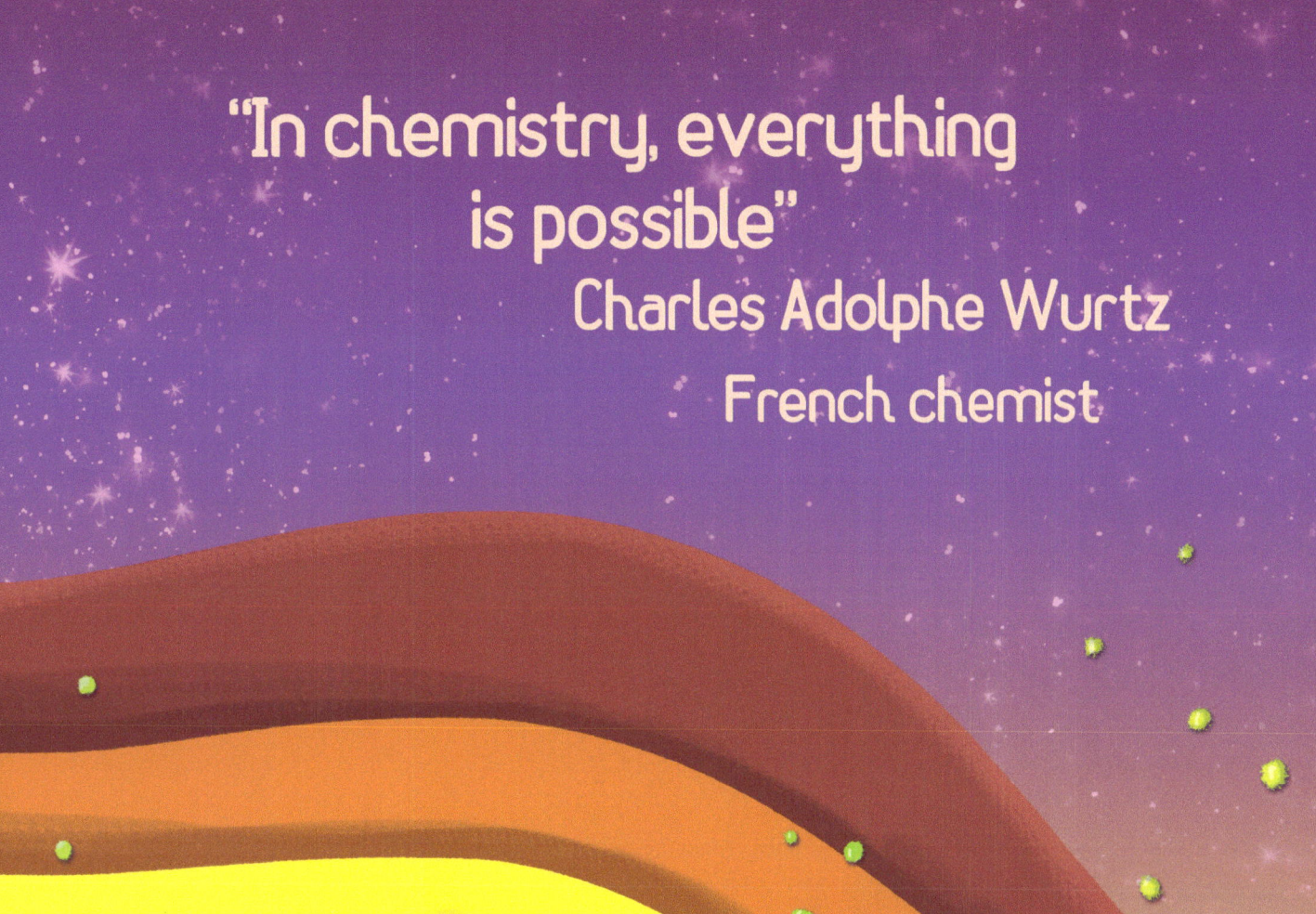

Chemistry is a branch of science that studies what everything is made of and how it works. Ice formation is chemistry. Medicines and paints are all chemistry. What is chemistry? In this book, we will find out!

We are all chemists. Chemistry is part of everything in our lives. Chemistry is involved in everything we do, from growing and cooking food to cleaning our homes and bodies to launching a space shuttle.

# FOOD CHEMISTRY

Food science deals with the three biological components of food — carbohydrates, lipids and proteins. Carbohydrates are sugars and starches the chemical fuels needed for our cells to function. Lipids are fats and oils and are essential parts of cell membranes and to lubricate and cushion organs within the body. Because fats have 2.25 times the energy per gram than either carbohydrates or proteins, many people try to limit their intake to avoid becoming overweight. Proteins are complex molecules composed of from 100 to 500 or more amino acids that are chained together and folded into three-dimensional shapes necessary for the structure and function of every cell.

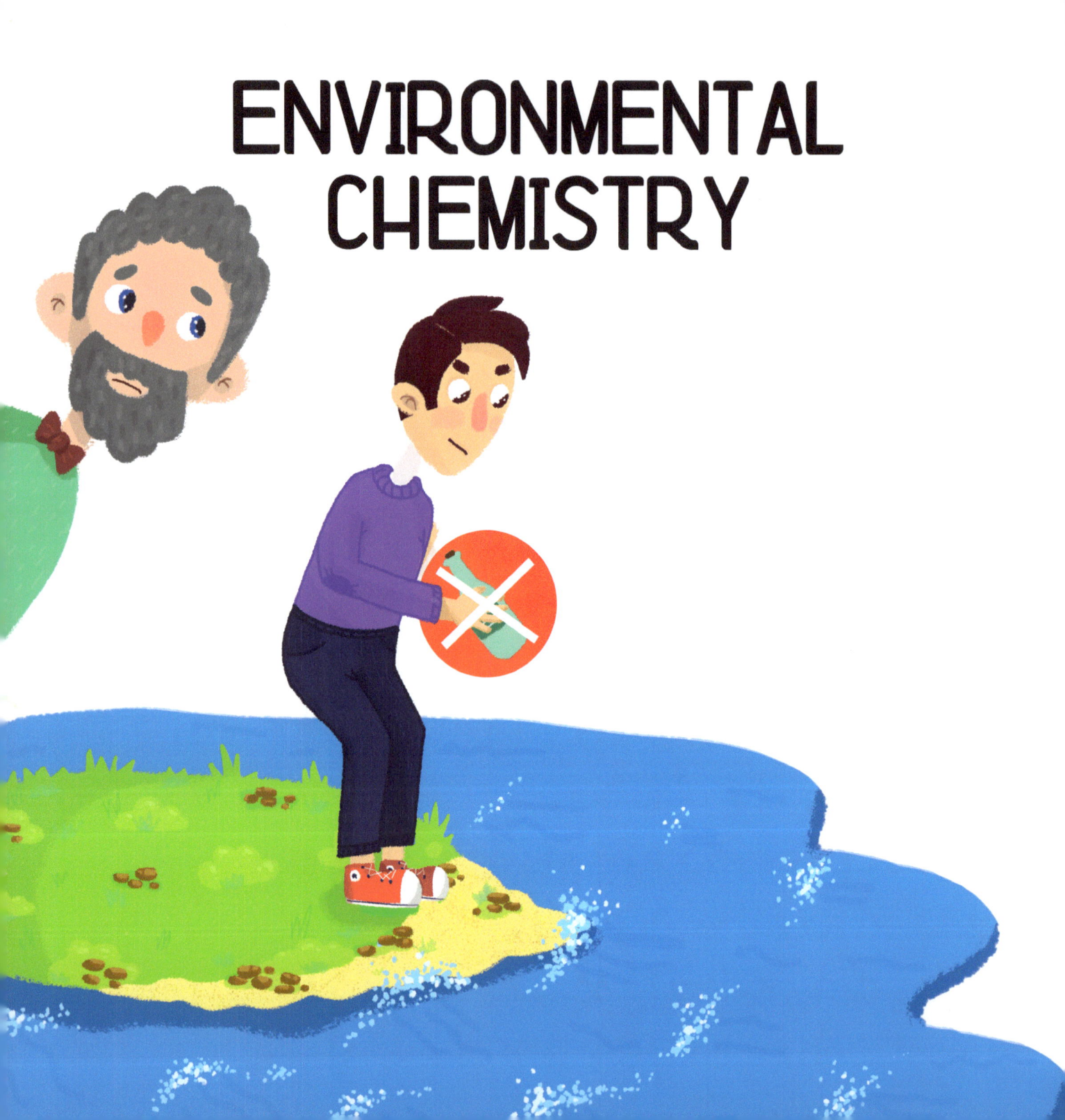

Environmental chemists study how chemicals interact with the natural environment. Analysis can determine if human activities have contaminated the environment or caused harmful reactions to affect it. Water quality is an important area of environmental chemistry. "Pure" water does not exist in nature; it always has some minerals or other substance dissolved in it. Water destined for human consumption must be free of harmful contaminants and may be treated with additives like fluoride and chlorine to increase its safety.

November 21, 1783 is a significant day in the history of aeronautics. On this day, two brave Frenchmen: Pilatre de Rozier and Marquis d'Arland flew for the first time in history in a balloon that belonged to the Montgolfier brothers. The balloonists reached an altitude of 915 m, as if 150 giraffes stood on top of each other. The brothers Joseph and Jacques-Étienne had the idea to make their own unique invention - a balloon that could cover distances.

They were led to this decision by numerous studies of various chemists and physicists.

In 1766, after the discovery, Henry Cavendish (chemist) found out that the "combustible air" in density is several times less than the air itself. The Montgolfier brothers decided to conduct their experiments, filling the shirt with air hot from the fire, and then paper bags. Further numerous tests were carried out launching balls of silk and flax. Stuffed items rose to the ceiling, which was already a huge break through.

Facts about Marie Curie:
- Marie became the Professor of Physics at the Sorbonne. She was the first woman to hold this position.
- Marie became good friends with fellow scientist Albert Einstein.
- Her first daughter, Irene, won a Nobel Prize in Chemistry for her work with aluminum and radiation.
- Marie had a second daughter named Eve. Eve wrote a biography of her mother's life.
- The Curie Institute in Paris, founded by Marie in 1921, is still a major cancer research facility.

# NEW ELEMENTS

Marie and her husband spent many hours in the science lab investigating pitchblende and the new element. They eventually figured out that there were two new elements in pitchblende. They had discovered two new elements for the periodic table! Marie named one of the elements polonium after her homeland Poland. She named the other radium, because it gave off such strong rays. The Curies came up with the term "radioactivity" to describe elements that emitted strong rays.

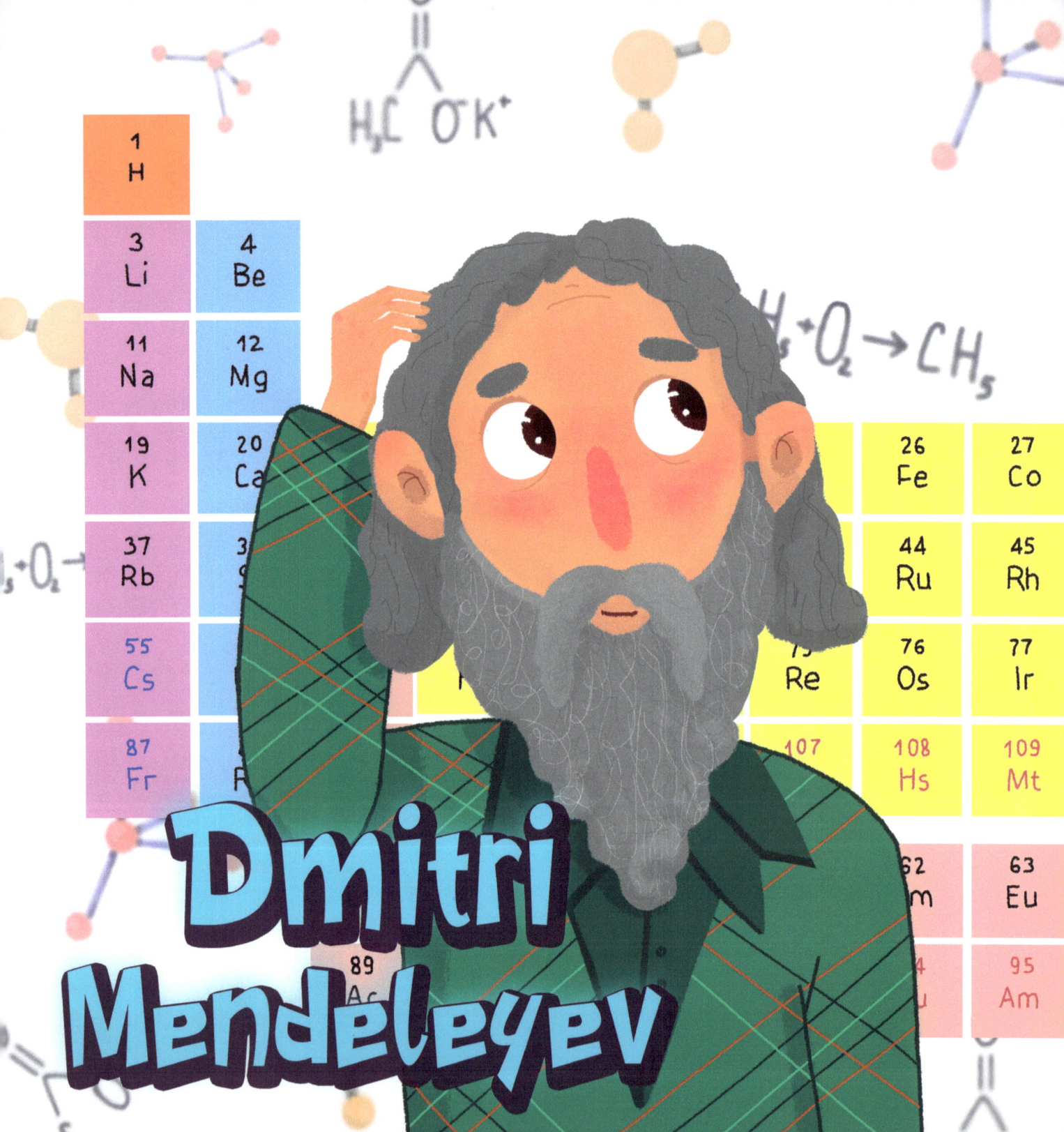

Dmitri Mendeleyev was a Russian chemist who came up with the first periodic table of the elements which he published in 1865. He was able to predict the discovery of many more elements using the table.

# Alfred Nobel

Alfred Nobel was a Swedish chemist and inventor who invented dynamite. He was a prolific inventor and held 350 patents. He is perhaps most famous for starting the Nobel Prize. The element nobelium is named after Alfred Noble.

The atom is the basic building block for all matter in the universe. Atoms are extremely small and are made up of a few even smaller particles. The basic particles that make up an atom are electrons, protons, and neutrons. Atoms fit together with other atoms to make up matter. It takes a lot of atoms to make up anything. There are so many atoms in a single human body trillions and trillions and trillions.

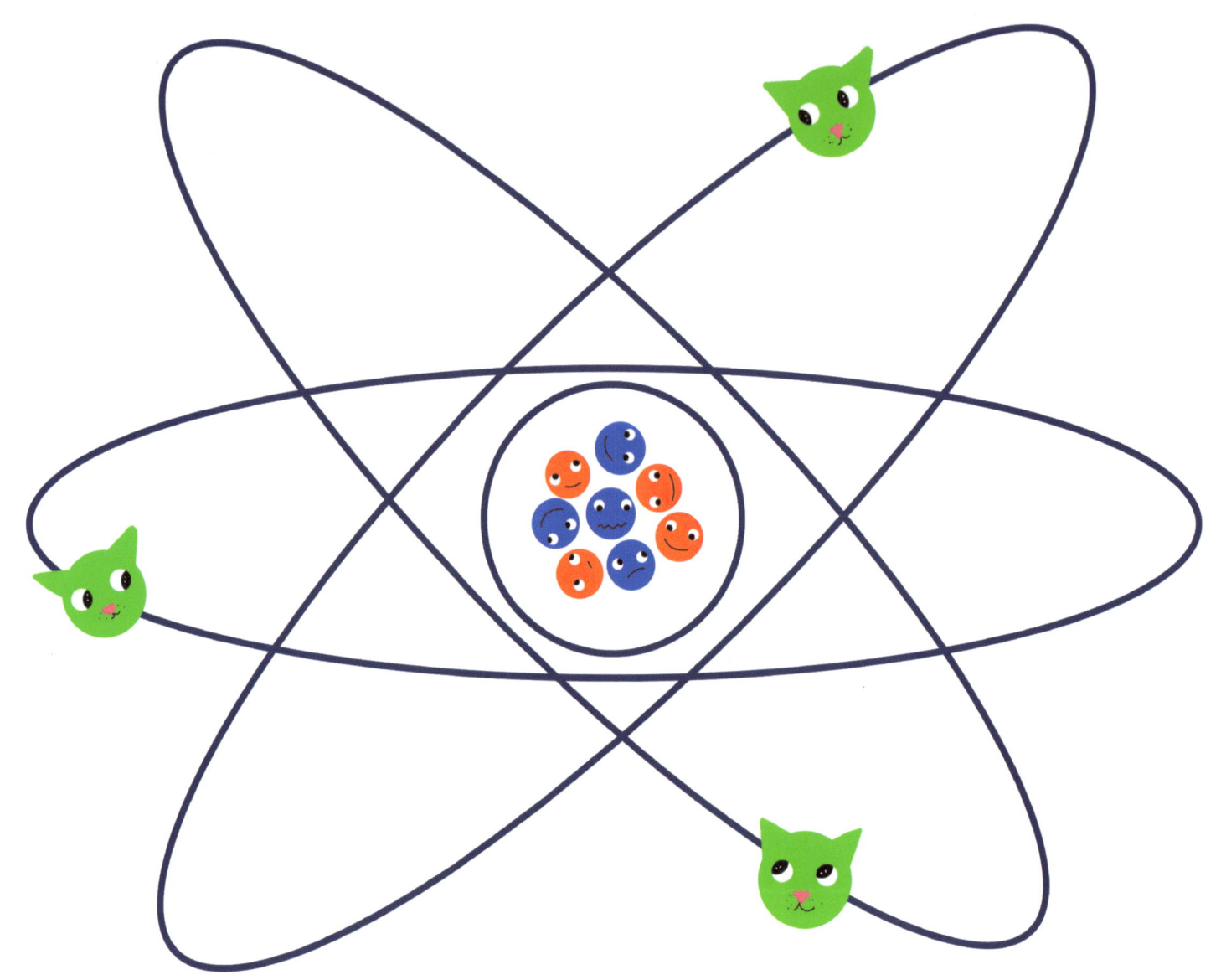

# Structure of the Atom

At the center of the atom is the nucleus. The nucleus is made up of the protons and neutrons. The electrons spin in orbits around the outside of the nucleus.

The proton is a positively charged particle that is located at the center of the atom in the nucleus.

# The Proton

The electron is a negatively charged particle that spins around the outside of the nucleus. Electrons spin so fast around the nucleus. If there are the same number of electrons and protons in an atom, then the atom is said to have a neutral charge.

Electrons are attracted to the nucleus by the positive charge of the protons. Electrons are much smaller than neutrons and protons. About 1800 times smaller!

# The Electron

# The Neutron

The neutron doesn't have any charge.
The number of neutrons affects the mass
and the radioactivity of the atom.

HOLZ books | 2018

# www.holzbooks.com

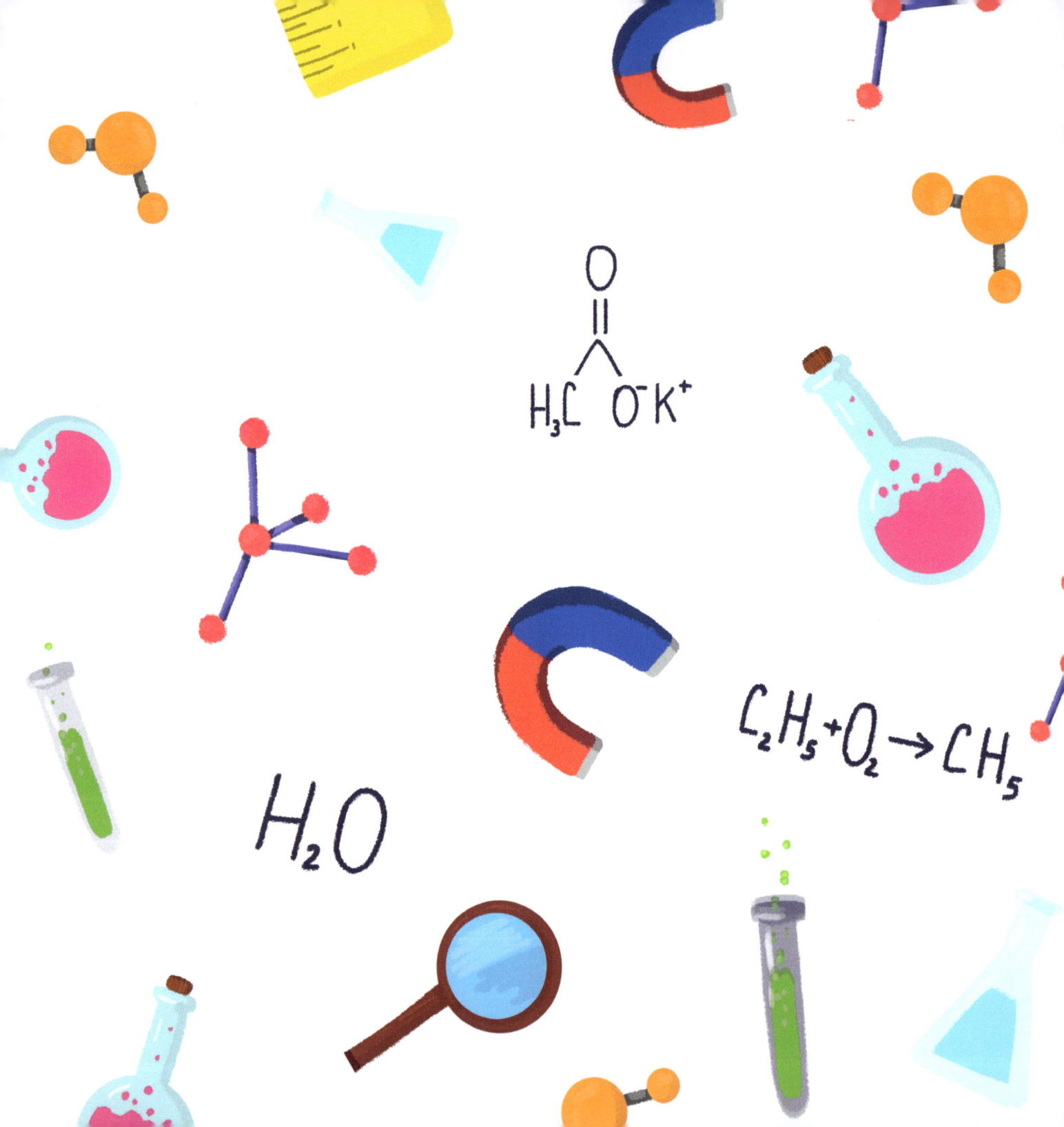

www.ingramcontent.com/pod-product-compliance
Lightning Source LLC
Chambersburg PA
CBHW041933240526
45473CB00034B/1232